A PEEK AT LEAVES

DOUG BRADLEY

Published in 2026 by The Rosen Publishing Group, Inc.
2544 Clinton Street, Buffalo, NY 14224

First Edition

Editor: Greg Roza
Book Design: Leslie Taylor

Photo Credits: Cover, p. 1 Pakhnyushchy/Shutterstock.com; p. 5 brgfx/Shutterstock.com; p. 7 (maple) djgis/Shutterstock.com, (basil) Anna Klymchuk/Shutterstock.com, (clover) Larbi chelh/Shutterstock.com, (palm) Ken stocker/Shuttertock.com; p. 9 (top) Cora Mueller/Shutterstock.com, (middle) Kati Maria/Shutterstock.com, (bottom) Jerry photo/Shutterstock.com; p. 11 Designer things/Shutterstock.com; p. 13 Emvat Mosakovskis/Shutterstock.com; p. 15 ozkan ulucam/Shutterstock.com; p. 17 (top) barmalini/Shutterstock.com, (bottom) josehidalgo87/Shutterstock.com; p. 19 Vasit Buasamui/Shutterstock.com; p. 21 Sinn P. Photography/Shutterstock.com.

Library of Congress Cataloging-in-Publication Data

Names: Bradley, Doug, 1971- author.
Title: A peek at leaves / Doug Bradley.
Description: [Buffalo] : PowerKids Press, [2026] | Series: A peek at plants
  | Includes bibliographical references and index.
Identifiers: LCCN 2024055788 (print) | LCCN 2024055789 (ebook) | ISBN
  9781499452013 (library binding) | ISBN 9781499452006 (paperback) | ISBN
  9781499452020 (ebook)
Subjects: LCSH: Leaves–Juvenile literature. | Plants–Juvenile literature.
Classification: LCC QK649 .B73 2026 (print) | LCC QK649 (ebook) | DDC
  581.4/8–dc23/eng/20250107
LC record available at https://lccn.loc.gov/2024055788
LC ebook record available at https://lccn.loc.gov/2024055789

Manufactured in the United States of America

CPSIA Compliance Information: Batch #CSPK26. For Further Information contact Rosen Publishing at 1-800-237-9932.

Find us on 

CONTENTS

Plant Parts!

Plants can look very different from one another, but most have the same parts. Roots grow down into the ground. Stems grow up into the air. Flowers and fruits help a plant to grow seeds. A plant's leaves have a very special job. Let's learn all about leaves!

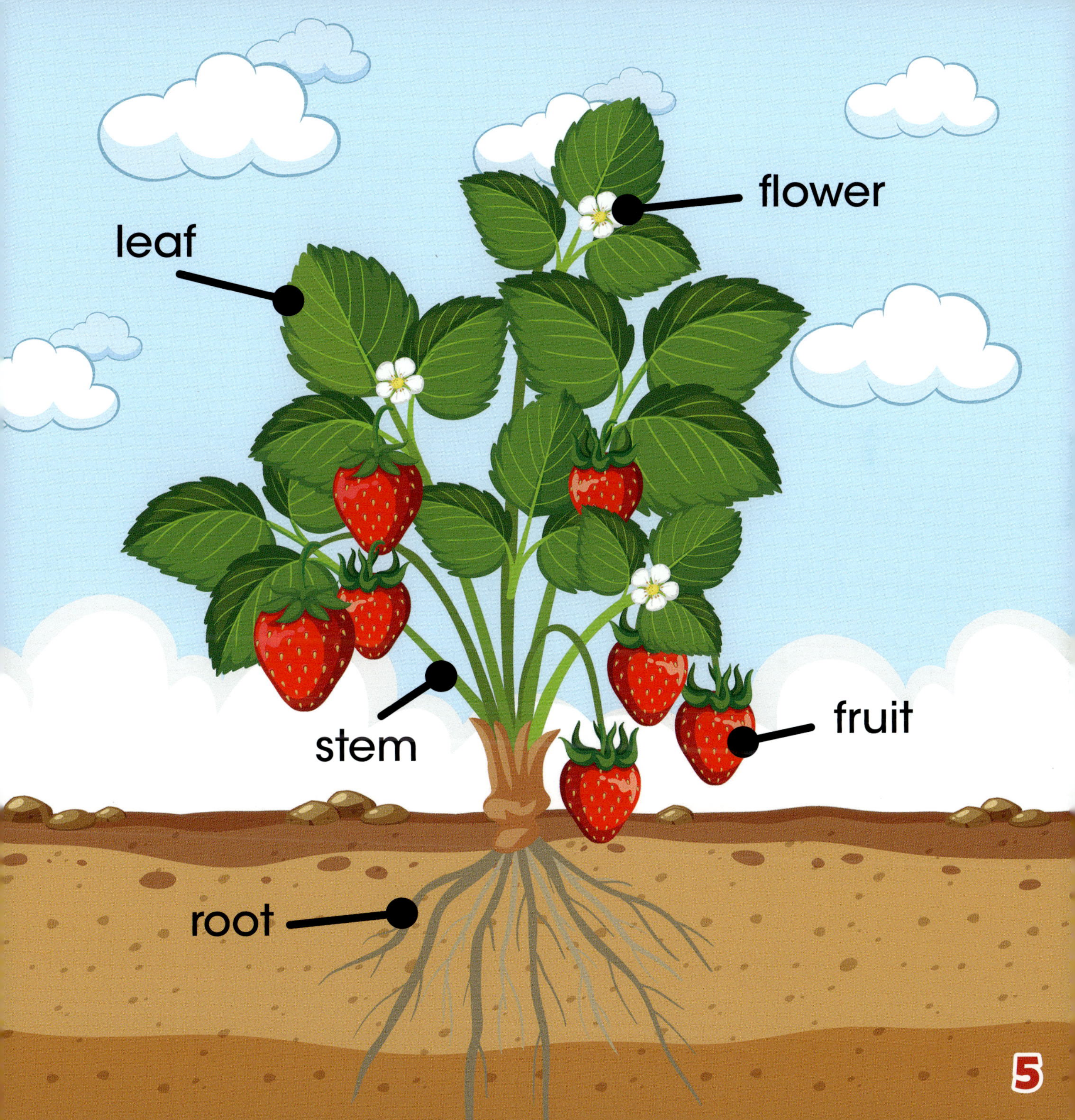

flower
leaf
stem
fruit
root
5

Shape and Size

Leaves come in many shapes and sizes. Some have even edges. Some have uneven edges. Most leaves are flat. The largest leaves grow on palm trees. The smallest leaves are so small they are hard to see! As we will soon learn, all leaves have the same job.

maple leaves
basil leaves
clover leaves
palm tree leaves

From Bud to Leaf

A leaf starts as a bud. Buds form on a plant's stem or a **branch**. Leaves grow larger thanks to the plant's roots! Roots take in water and **nutrients** from the ground. These things pass into the stem and then into the leaves through very tiny **tubes**.

9

Making Food

Leaves are green. This green color is made by something inside the leaves that takes in sunlight. Leaves use sunlight and a gas called carbon dioxide to make sugar. Plants use sugar as food. Plants also give off a gas called **oxygen** during this action, which is called photosynthesis.

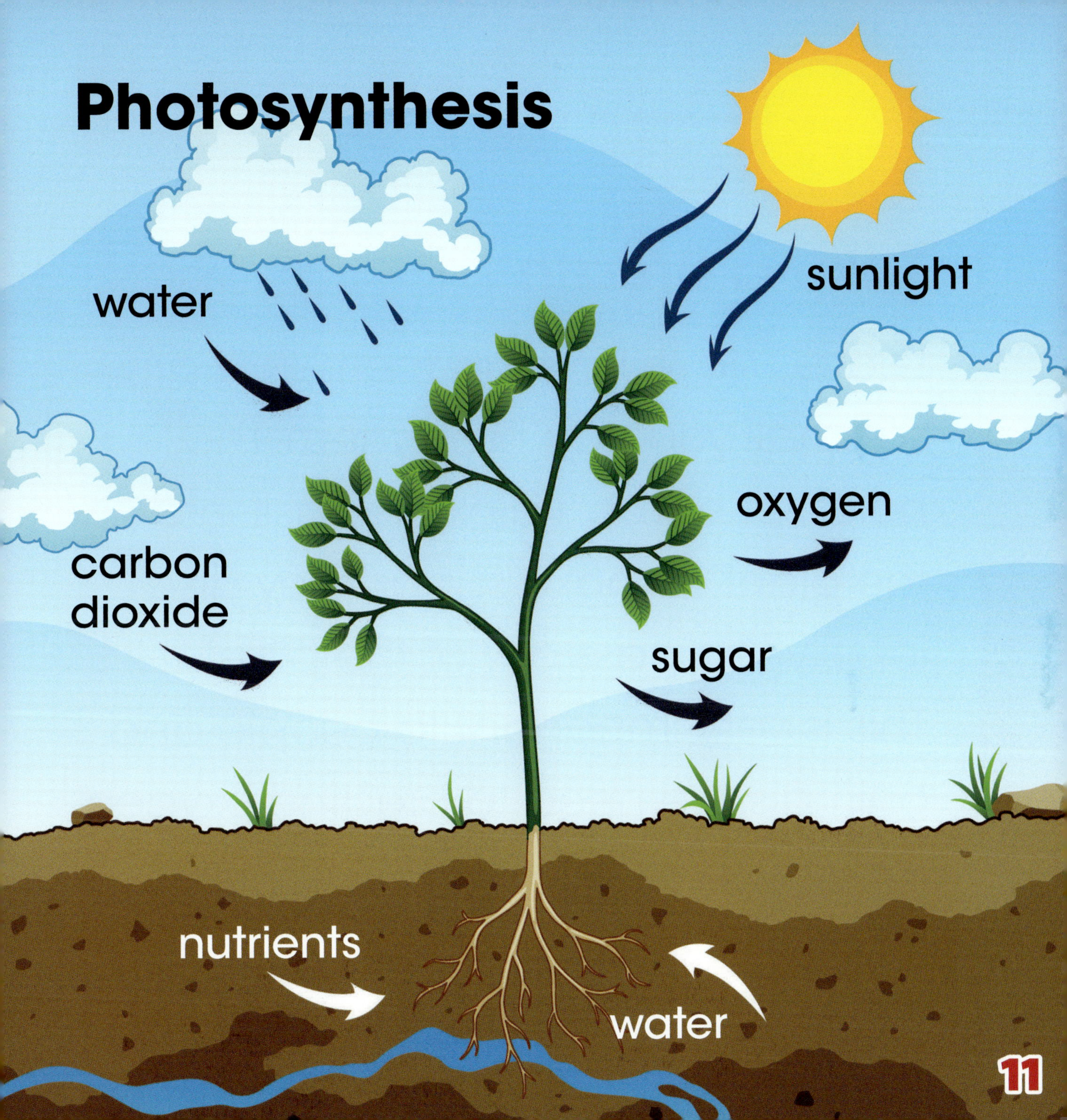

Photosynthesis
water
sunlight
carbon
dioxide
oxygen
sugar
nutrients
water
11

Take a Breath

Leaves have holes in them that are so small you can't see them. These holes let carbon dioxide pass into the plant to make sugar. They also let oxygen pass out of the plant. This is very important to people and animals. We need to breathe oxygen to live!

Changing Colors

In places that have all four seasons, most leaves change color. This is because there's not as much sunlight at this time of year. Leaves start to lose their green color. They turn brown, yellow, red, orange, and even purple! Soon, the leaves fall to the ground.

15

Is That a Leaf?

Some leaves don't look like leaves at all! Pine trees have leaves that are long and pointy. Pine trees stay green all year long, so they are called evergreen trees. The **spines** on cactus plants are leaves. The Venus flytrap has leaves that trap and eat flies!

cactus
Venus flytrap

The Use of Leaves

You may have a salad with lettuce, spinach, or another type of leaf. But there are other uses for leaves. Leaves are used to make **medicines**. Some leaves keep bugs away. Some people use long, thin palm leaves to make objects, such as baskets and hats!

Grow Your Own Leaves!

You can start your own garden and watch leaves grow. In time, you can even pick some leaves and eat them! A salad made with home-grown lettuce is tasty! If you live where leaves change color in the fall, you can collect them and use them for crafts!

GLOSSARY

branch: A smaller part of a plant's stem that grows off the stem and often grows leaves, fruits, and flowers.

medicine: A drug that a doctor gives you to help fight illness.

nutrient: Something taken in by a plant or animal that helps it grow and stay healthy.

oxygen: A gas in the air that plants make and animals need to breathe to live.

spine: A sharp, pointed part of a plant, often found on its stem or branches.

tube: A long, thin channel within a plant or animal that moves water and nutrients.

FOR MORE INFORMATION

BOOKS

Maloney, Brenna. *Leaves*. New York, NY: Children's Press 2025.

Silvora Collections. *Plant Parts and their Purpose: A Budding Botanist's Activity Book*. Independently published, 2024.

WEBSITES

Photosynthesis for Kids
www.youtube.com/watch?v=lIn136eMl4g
Watch this short, fun video to learn more about the process of photosynthesis.

Why Do Leaves Change Color?
scijinks.gov/leaves-color
Have you ever wondered why leaves change color? Read more about it here.

INDEX